THE AUSTRALIAN NATIONAL UNIVERSITY

SDSC

Strategic & Defence Studies Centre

Working Paper No. 412

Keeping our Yellowcake Peaceful: A Policy Framework for Uranium Exports

Ron Huisken

March 2009

Dr Ron Huisken joined the Strategic and Defence Studies Centre at the Australian National University in 2001, after nearly twenty years working in the Australian government departments of Foreign Affairs & Trade, Defence, and Prime Minister & Cabinet. His research interests include US security policies, multilateral security processes in East Asia, alliance management and non-proliferation. Dr Huisken has authored numerous works, including a number of working and Canberra papers published by the SDSC. This paper represents the author's views alone. It has been drawn entirely from open sources, and has no official status or endorsement.

This paper was first prepared for the uranium workshop at Flinders University, 7–8 June 2007. It is reprinted here, with amendment.

Published by the Strategic and Defence Studies Centre at The Australian National University, Canberra

National Library of Australia Cataloguing-in-Publication entry

Author: Huisken, R. H. (Ronald Herman), 1946—

Title: Keeping our Yellowcake Peaceful: A Policy Framework for Uranium Exports

Edition: 1st ed.

ISBN: 9780731554881 (pbk.)

Series: Working Paper (The Australian National University. Strategic and Defence Studies Centre); no. 412

Subjects: Uranium industry—Government policy.
Nuclear energy—Government policy.
Nuclear nonproliferation—Government policy.
Nuclear weapons—Government policy.

Dewey Number: 338.274932.

Strategic and Defence Studies Centre's Publication Program

Established in 1966, the SDSC is located within the Research School of Pacific and Asian Studies at The Australian National University, Canberra, Australia. It specialises in the study of strategic issues—predominantly in the Asia-Pacific region.

The Canberra Papers on Strategy and Defence series is a collection of monograph publications arising principally from research undertaken at the SDSC. Recent previous Canberra Papers have focused on major aspects of Australian defence policy, disease security in Northeast Asia, ballistic missile defence, cyber-warfare, the complexities of dealing with radical Islam, and aspects of transnational crime.

As at 2006, all papers in the Canberra Papers on Strategy and Defence series are peer reviewed.

In addition, the SDSC also holds a number of 'one-off' publications.

SDSC Working Papers are shorter than Canberra Papers and focus on areas of current research interest to SDSC academic staff or the Centre itself. Topics of previous Working Papers have included Australia's defence policies, Australia's security relationship with Japan, intelligence on Iraq's WMD, Western air power, long-range missiles, instability in the US-ROK alliance, the balance of power in East Asia, New Zealand's defence policies, aspects of future land warfare, and the threat of terrorism on regional development.

Most working papers published between 2003 and 2009 can be downloaded for free from the SDSC publications webpage at <http://rspas.anu.edu.au/sdsc/publications.php>. This page also lists those Centre publications in print that are available for purchase, together with an ordering form.

Publisher

Strategic and Defence Studies Centre
Australian National University
Canberra, ACT 0200, Australia

Tel: +61 2 6125 9921
Fax: +61 2 6125 9926
Email: http://rspas.anu.edu.au/sdsc

Keeping our Yellowcake Peaceful: A Policy Framework for Uranium Exports

Ron Huisken

Accumulating concerns about greenhouse gas emissions, global warming and climate change have resuscitated interest around the world in nuclear power generation. Australia has both joined those taking a more positive look at nuclear power generation, and, as a key player in the global market for uranium, begun to think about how best to capitalise on this revival of global interest. But uranium is a very special commodity. A relatively modest quantity of uranium can yield prodigious amounts of energy at a controlled rate over long periods of time, but it can also be processed so that the composition of natural uranium (0.7 per cent energy-producing U235 plus 99.3 per cent inert U238) is roughly reversed, in which case it can yield a prodigiously destructive bomb. Further, a controlled chain-reaction in a uranium-fuelled reactor produces plutonium as a by-product, and plutonium is the alternative fuel for the bomb. In other words, it all starts with uranium, and Australia owns an estimated 40 per cent of the world's known reserves, making us the Saudi Arabia of the nuclear business.

Australia decided early that it would be smart to make our way without nuclear weapons, but it recognised that this would only remain a smart (or, indeed, viable) road if essentially everyone also decided to live without the bomb. Accordingly, Australia became a strong and consistent champion of nuclear non-proliferation and has used the regime that has been developed to accomplish this objective as the framework within which it can export uranium responsibly: Australia will only sell uranium to countries that are party to the Non-Proliferation Treaty (NPT), that have full-scope safeguards agreements with the International Atomic Energy Agency to allow the non-military intent of their nuclear activities to be verified and, more recently, that are prepared to conclude a bilateral agreement that provides some further assurance that Australia's uranium will not contribute to a bomb program.

Unfortunately, the reassurance provided by the non-proliferation regime has weakened very considerably over the past 10–15 years. Some would contend that we should not be surprised by recent instances of nuclear proliferation. These folk, who label themselves Realists, proceed from the fact that the bomb cannot be uninvented, and from the assessment that it is a decisively important strategic tool, to the view that the only sensible perspective on nuclear weapons is that they will spread and that the challenge has always been and will always be to cope with this diffusion. Even Realists, however, recognise that an epidemic of proliferation could readily overtax our capacities to cope.

Those who sense that confidence in our capacity to cope with more nuclear weapons in the hands of more states might be misplaced must also deal with the fact that the bomb cannot be uninvented. A commitment to non-proliferation should be informed by the fact that the challenge is not defined by the proliferation risks at any point in time but constitutes a challenge that will exist in perpetuity. We have already been through a phase in which two powerful rivals lost touch with reality and ended up in a ridiculously dangerous situation where each could assuredly destroy the other as a functioning society and, just possibly, put a question mark on the survival of our species: not global warming but a nuclear winter. While that particular concern has receded, the risk of a revival, with the same or different players, will exist in perpetuity. It is worth noting that the present non-proliferation regime was put in place essentially because these two powers became sufficiently alarmed that some third nuclear force could derail the 'delicate balance of terror' between them.

The nuclear non-proliferation enterprise has always recognised that the scientific, engineering, and industrial 'secrets' of the bomb would seep out and that, over time, an increasing number of countries would develop the wealth and the skills to make acquiring the bomb a realistic proposition. In other words, it was recognised from the outset that there would be an inexorable secular decline in the efficacy of supply-side controls (restricting access to bomb-related materials and technology) and that progressively greater reliance would have to be placed on voluntary abstinence or suppressing the demand for this capability (that is, a state consistently deciding that the balance of its interests lay in remaining a non-nuclear weapon nation).

The nuclear non-proliferation regime has lost valuable coherence and credibility because both the supply and demand sides of the equation have taken some heavy hits. Iraq, for example, was a party to the NPT and subject to regular inspections by the International Atomic Energy Agency (IAEA) during the 1970s and 1980s, but when UN inspectors entered the country in 1991 following the first Gulf War they found an array of bomb-related nuclear programs that exceeded the worst suspicions of the major intelligence agencies around the world and which had been progressing under the noses of the IAEA inspectors. The IAEA had been required to accept at face value whatever a state declared to be the extent of its nuclear activities. It could not question a declaration on the basis, for example, of reports in the trade literature, or diplomatic scuttlebutt, or intelligence provided by another state. The Iraqi case proved sufficiently blatant and embarrassing to permit the IAEA to design a declaration and inspection process—the so-called Additional Protocol—that did not rely so completely on the good faith of member states. The Additional Protocol, while not the new compulsory standard, is steadily gaining voluntary adherents.

More than a decade later, Libya's surprise 2003 announcement of a deal with the United States and the United Kingdom to verifiably terminate its several clandestine Weapons of Mass Destruction (WMD) programs provided hard evidence of a black market in sensitive nuclear technologies centred on Dr A.Q. Khan, the father of Pakistan's bomb program.[1] The unravelling of this network created something akin to a perfect storm for the likes of Iran and North Korea. Iranian deception of the IAEA regarding uranium enrichment and plutonium had been exposed by a dissident group late in 2002. Tehran's efforts to limit this exposure were demolished by A.Q. Khan's revelations, compounding perceptions that Iran had indeed embarked on a clandestine weapon program. Similarly, Khan's confessions confirmed some Pakistan-North Korea dealings that exacerbated the revived nuclear crisis on the Korean Peninsula. The major intelligence agencies had known of this network in broad terms for a number of years, but had been unable to penetrate it. The eye-opening aspect of the Khan network was that it was a robust and reliable means of acquiring some of the pivotal technologies for the bomb. The centrifuges for the enrichment of uranium intercepted while being shipped from Malaysia to Libya included components from some nine countries, including the likes of South Korea, Japan and Switzerland. Moreover, while the centrifuges from Malaysia were removed, the same vessel carried a second shipment of centrifuges for Libya assembled in a Turkish facility and these were not detected until they were unloaded in Libya and handed over under Gaddafi's deal with the United States and the United Kingdom. Meanwhile, the blueprint for a basic uranium-based bomb, provided to Pakistan by China (probably in the 1980s), also turned up in Libya and, it would seem, at least partially, in Iran.

The developments on the demand front were a good deal more subtle and nuanced, but these too both reflected and reinforced a weakening of the norm against the possession of nuclear weapons, resulting in growing dissonance and friction over the enforcement and strengthening of the non-proliferation regime.

The end of the Cold War and the demise of the Soviet Union inevitably saw additional responsibilities and stronger expectations for leadership drift toward Washington (the Gulf War in 1991, the North Korean nuclear crisis of 1991–94, the several phases (1994–99) of the break-up of the former Yugoslavia, and the clamouring of states of the former Soviet empire for new economic and security ties to western Europe and NATO.) By and large, over the decade of the 1990s, despite its new prominence and exposure as the most powerful state since the Roman Empire, the United States succeeded in protecting its image of a reluctant, more or less benign and thereby thoroughly acceptable hegemon. At the same time, it led visibly to escalating concerns in Washington about the risks and vulnerabilities that seemed to accompany this status in the new world order. Foremost among these were the proliferation of WMD and long-range missile systems to so-called 'rogue states', and terrorism.

Against this background, counter-proliferation steadily became a more prominent strand in US non-proliferation policy. Many observers in the late 1980s and early 1990s were inclined to the view that the non-proliferation regime was indeed breaking down. US intelligence assessment tended to confirm these impressions by anticipating the continued spread of WMD and long-range missiles. Counter-proliferation, quite obviously, amounts to a 'the horse has bolted' posture. The intent behind counter-proliferation is threefold: (1) to devalue the possession of WMD through the acquisition of military capabilities that can destroy these weapons (and the means to make them) before they can be used; (2) to develop better defences against these weapons (missile defences, better protective clothing, immunisation and so on); and (3) to take steps to minimise the damage and disruption that would result from the use of WMD.

The Clinton Administration added active defences (including theatre missile defences) consistent with the Anti-Ballistic Missile (ABM) Treaty to its policy mix, but retained the prevention of proliferation through the Treaty regime as its primary instrument. Additionally, however, powerful signals of the growing caution or pessimism on this front within the wider US security community came when the US Congress refused to endorse ratification of the Nuclear Test Ban Treaty that had been negotiated in 1996, and the fierce pressure from Republicans to pursue missile defences as hard and as far as US technology could take them.

The geopolitical thinking that accompanied President George W. Bush into office was the antithesis of 'soft and reluctant hegemony', a mindset that was turbo-charged by the dreadful terrorist attacks on 11 September 2001. The message from Washington, eventually starkly amplified through the doctrine of pre-emption and Operation *Iraqi Freedom*, was that US policy on the proliferation of WMD had been sharply telescoped or simplified. In the past, a proliferation issue would be addressed through assessing motives and trying to devise a mixture of incentives and penalties to constrain the state concerned. Washington was now saying that its willingness to manage proliferation risks was much weakened. When deemed necessary—especially when it perceived a current or possible future risk that WMD will be shared with terrorists—Washington would require that the proliferation be comprehensively and permanently reversed.

Washington has since moved energetically to inject a new seriousness of purpose into non-proliferation: to require that the non-proliferation regime be rigidly policed; that a clear and consistent way of imposing consequences on states that pass certain milestones be developed; and that non-proliferation objectives be upheld regardless of the temptations to trade them off against economic or political interests. In this regard, Washington has:

- substantially increased funding to the Nunn/Lugar Threat Reduction Initiative from the early 1990s to make secure and/or destroy WMD stocks in Russia;
- inspired the G8 in June 2002 to form the Global Partnership Against the Spread of Weapons and Materials of Mass Destruction (with a target budget of US$20 billion over 10 years);
- pressed (unsuccessfully) to have the so-called Additional Protocol, a more comprehensive and intrusive IAEA inspection regime, adopted as the global standard at the NPT Review Conference in 2005;
- pressed complementary instruments of control—the Australia Group, Nuclear Suppliers Group, Zangger Committee, Missile Technology Control Regime, Wassenaar Arrangement—to toughen their guidelines on permissible commerce;
- launched the Proliferation Security Initiative (in May 2003) which seeks to put in place the ability to interdict the international transfer of WMD and long-range missiles by countries like North Korea; and
- been a prime mover behind UN Security Council Resolution 1540 (adopted unanimously in April 2004), which imposes a legal obligation on states to establish and enforce laws to prevent the proliferation of WMD (and precursor materials and technology) and their means of delivery.

Except for the first two points, all these initiatives address tougher enforcement of the existing regime, rather than transformation of the regime itself.

The heightened dissonance in the international arena referred to earlier was echoed in the specific field of nuclear weapons and proliferation. It became ever harder to discern what was driving particular US policy settings—the 'war on terror' or the distinctive worldview the administration had begun to unveil before the terrorist attacks on New York and Washington. Inevitably, a number of key players began to see the 'war on terror' as a vehicle the administration was exploiting to advance its deeper geopolitical agenda.

A key indicator of Washington's new mindset was its single-minded determination in 2001 to scrap the ABM Treaty. The Bush Administration had two primary objectives in doing this.

First, it wanted to give American technology an unrestricted environment in which to pursue the objective of effective ballistic missile defences. It did seek to reassure Russia and China, in particular, that its ballistic missile defence (BMD) objectives were confined to the defence of US and allied forces in distant theatres and to deter/defeat the (very modest) threat that a 'rogue' state like North Korea might at some point pose to the continental US itself. Second, the Bush Administration wanted to complete the decoupling of Russia (as the successor to the Soviet Union) from the United States as a co-manager of strategic stability, including negotiated reductions in strategic offensive forces. Washington's preference, naturally enough, was that Moscow would agree that the ABM Treaty was an anachronism that they could jointly consign to the dustbin of history. In the event, Washington had to unilaterally give notice of its intention to withdraw, which it did in December 2001. In terms of the dynamics of strategic nuclear relationships, the demise of the ABM Treaty meant that Russia and China now had only Washington's political assurances that the United States did not intend its BMD capabilities to deny either country a nuclear deterrent capacity against the United States.

In a similar vein, Washington's Nuclear Posture Review (NPR), copiously leaked in January 2002, firmly re-established nuclear weapons as a central component of America's security posture, including the intent to maintain a force that was emphatically superior to all others, and capable of being rapidly augmented. The review exposed US plans to reduce to some 2000 operationally available strategic nuclear warheads by 2012, but to have some 2400 additional warheads available for re-deployment at relatively short notice.

This nuclear posture was proclaimed to be the first that had been shaped without regard to balancing the forces of a peer competitor. It represented America's unilateral requirement for strategic nuclear forces, being the smallest possible force it needed. The additional message in this posture, and a critically important one in the non-proliferation context, was that the United States seemed to be ruling out further reductions more or less indefinitely. Alongside abrogation of the ABM Treaty, Washington insisted that it would henceforth determine its nuclear force requirements unilaterally, not jointly with Russia (or anyone else) in the mould of the Strategic Arms Limitation Talks (SALT) and Strategic Arms Reduction Treaty (START) agreements. It ultimately relented to the point of concluding the Moscow Treaty in May 2002, but insisted that the text of the treaty basically comprise the postures that each party had determined unilaterally.[2]

The NPR also signalled the intention to streamline the process of planning and preparing options for the use of nuclear weapons, to get away from the elaborate and rigid strike options of the past, and to make the nuclear forces more flexible and responsive to whatever 'extreme contingencies' may arise. In this vein, it kept alive the option of developing a new nuclear weapon tailored to the counter-proliferation mission—a low-yield, earth-penetrating 'bunker buster'. Such a weapon would be an additional deterrent to future proliferators hoping to secure critical nuclear facilities underground, but for the current crop of 'wannabes' it would be intended for use rather than deterrence.

The Bush Administration also scrupulously avoided any reference to the pursuit of nuclear disarmament as a long term goal or as an obligation under Article VI of the NPT. The last occasion on which the United States embraced the spirit of Article VI of the NPT was the Review Conference in May 2000 when the Clinton Administration was still in power. At that time, the five recognised nuclear weapon powers reaffirmed their 'unequivocal undertaking' to eliminate their nuclear arsenals. Statements by Bush Administration officials later confirmed that nuclear disarmament was not an objective that the United States was prepared to subscribe to at the present time: the typical formulation has been that the United States is 'moving in the direction required by the NPT'. In a message in April 2003 to the Preparatory Committee for the 2005 NPT Review Conference, then US Secretary of State Colin Powell said only that: "The Moscow Treaty and other US actions are based on a desire and an intention to reduce our reliance on nuclear weapons and to eliminate surplus stocks of weapons-grade material."[3]

A final important consideration concerns conventional military power. The Bush Administration, in a manner of speaking, pulled the plug on the accumulated technological developments (especially precision targeting at all ranges and in all weather and the manipulation and dissemination of information) that, when harnessed to organisational, doctrinal and tactical changes had long been labelled a (potential) Revolution in Military Affairs (RMA). The result was a qualitative great leap forward in US conventional military capability, quite literally 'redefining war on our terms', as Bush predicted shortly after his inauguration in 2001. This preponderance in demonstrably usable military power is thought by many to have intensified interest in nuclear weapons in states that fear preventive regime change, especially, of course, North Korea and Iran.

The combined effect of these developments—the erosion of confidence in the nuclear non-proliferation regime since around 1990 (Iraq, North Korea, the Indian and Pakistani tests, A.Q. Khan) and then the intersection of the reaction to the terrorist attacks of 11 September 2001 and the neo-conservative thesis that the United States could and should now overtly declare its intent to use (and, of course, protect) its unipolar status to project its values and interests more compellingly—has been to throw global governance into a tailspin. Even the spectacular global coalition that gathered spontaneously around Washington immediately after the attacks could not arrest the tide, and it withered away within a few months. From our non-proliferation perspective, the clearest indicator of the dissonance and loss of cohesion was that, just 10 years after the NPT had been extended indefinitely, the 2005 Review Conference ended in abject failure. Further evidence continues to accumulate, with the agonising difficulty of presenting North Korea and Iran with clear evidence of unity and resolve to enforce and protect the non-proliferation regime.

Perhaps it is time to concede the inevitability of the further proliferation of nuclear weapons and focus on coping strategies. The Realists, after all, have a point. Looking at the first 60 years of the nuclear era, the bomb may seem all but unusable (either because of retaliation in kind or because its destructiveness has always looked disproportionate to the politico-military interests it might advance). Still, whether possession of the bomb in an antagonistic relationship is symmetrical or asymmetrical, it has a decisive impact on how the rivalry plays out. If acquiring the bomb is feasible, or even imaginable, and states perceive that they have acute security concerns, they will crave the decisive edge it seems to provide (or at least deny that edge to their opponent).

Equally, however, all the impulses that have driven the non-proliferation enterprise to date remain in place. There still seems to be a universal 'gut feeling' that extending acceptance of the option to settle disputes through war to include use of nuclear weapons would be a major step backwards along civilisation's broadly positive trajectory. Everyone seems to be strongly attached to the norm (or taboo) on the use of nuclear weapons. There is, as noted earlier, the ever-present risk that nuclear use could spiral out of control and impose individual and collective costs of utterly irrational proportions. Additionally, there are now just nine states with nuclear weapons (the eight that have detonated a nuclear device plus Israel), while credible estimates suggest that some 40 states could implement a political decision to acquire the bomb. We may have lost some ground in recent times, but the floodgates seem to be holding. Another considerable virtue is that both the objective of non-proliferation and the failures have developed instincts to look harder and earlier into antagonistic relationships and seek to defuse them before ambitions to acquire nuclear weapons take root.

If these considerations leave the issue open, we now have in the al Qaeda phenomenon a further compelling reason to stick with non-proliferation. The new nightmare scenario, sharing the stage with the unauthorised use of nuclear weapons in the mould of Dr Strangelove, is a group of fanatics acquiring a bomb through theft or corruption and relishing its detonation in some urban conglomeration. No sophisticated strategic or political reasoning is called for here. The more places around our planet where nuclear weapons are made, repaired, stored and deployed, the higher the probability that somewhere, somehow, one or more of these weapons will fall into terrorist hands.

Assuming that the case for non-proliferation remains compelling, how can we halt and reverse recent trends and reinvest this enterprise with vitality, determination and coherence? A key step, in my view, is to set a new, specific non-proliferation objective, develop a comprehensive strategy to accomplish it and then consider the diplomatic modalities of combining the several elements of the strategy. For those who follow non-proliferation issues, the specific objective almost selects itself; namely, to get to the point where there is no national production of fissile material and where all fissile material is manufactured in internationally owned and operated facilities. In other words, bearing in mind that nuclear proliferation is a challenge that we must address in perpetuity, we should establish as the next major goal in the development of the international non-proliferation regime that all states surrender the right to have a complete nuclear fuel cycle under national control.[4]

This right was, of course, exercised by the five original nuclear powers before the global commitment to non-proliferation took shape, and by the handful of states that stayed outside the regime (most notably, Israel, India and Pakistan). It is also a right that is enshrined in the NPT, provided it is not exercised for military purposes. Non-nuclear weapon state parties to the NPT that have exercised the right to produce and stockpile fissile material include Germany, Japan, the Netherlands, North Korea (although it is debatable whether it was ever a full and genuine party) and, most recently, Iran.

A regime dedicated to precluding the acquisition of nuclear weapons, but which embraced five states that already had them, was naturally destined to be a stressful enterprise. These stresses were attenuated somewhat during the Cold War. That hegemonial struggle was spearheaded by a competition for nuclear supremacy and the international protest was focused more on the common danger of massive nuclear arsenals continuously on high alert than on the nuclear weapon states using their privileged possession of these weapons to ride roughshod over the interests of others. The safety valve of the Cold War disappeared in 1989–91 but, at least for a time, the visible dismantling of the postures for mutual assured destruction and, by the Cold War standards of nuclear arms control, reassuringly large and rapid reductions in the numbers of deployed weapons continued to mitigate the stresses that pervaded the non-proliferation regime. Among other things, it made possible the indefinite extension of the NPT in 1995. Since then, however, the challenges to the 'have/have not' nuclear regime, from both sides of the fence, have become relentless and cumulatively very damaging. We are still some distance from the situation where many states feel that prospects for proliferation are so strong that they should begin to position themselves to take this step more easily and quickly, but one senses that we are much closer to it than a decade ago.

The credibility of the proposition that the states with nuclear weapons are genuinely prepared to think about living without them lies at the core of the non-proliferation issue. This is such a huge proposition, and an objective that has seemed so remote since the early 1950s, that commitments to it have always had an air of 'convenient myth' about them. But myths can be important, especially convenient ones. The fact that no one knows whether nuclear disarmament can be achieved does not reduce the value of states making it plain that they want to take a hard look at this step and to do so from the most proximate vantage point that they can sensibly achieve. If non-nuclear weapon states cannot point to evidence of an interest in nuclear disarmament, they cannot be expected to resist indefinitely the increasingly easy option of acquiring nuclear weapons themselves.

Under the Bush Administration, the United States pursued policy imperatives that erased nuclear disarmament even as a convenient myth, accelerating the emergence of dissonance in responding to non-proliferation challenges and contributing, in the post-11 September 2001 atmosphere, to the posture of the unilateral use of force to preclude any nuclear-related capabilities in selected countries. Washington's policy settings also ensured a very hostile environment for its interest in a 'robust replacement warhead' (to replace the high-performance but high-maintenance designs that seemed a good idea during the Cold War when regular tests were possible), or in a low-yield, earth-penetrating weapon to shape the thinking of proliferators disposed to think that critical capabilities could be rendered invulnerable.

A primary task is to encourage the United States to reconsider its present settings and determine whether its non-proliferation interests into the indefinite future might be more securely advanced through developing the present regime. A rhetorical shift on article VI of the NPT, opening the door to ratification of the Comprehensive Test Ban Treaty (CTBT) and, perhaps most importantly, signalling a willingness to look beyond its present nuclear force structure with respect both to *numbers* and *readiness*, would all be means by which some measure of coherence could begin to be restored.[5]

A further core task is to galvanise the longstanding intent to negotiate a treaty banning the production of fissile material for military purposes, the so-called Fissile Material Cut-Off Treaty. The five recognised nuclear weapon-states are all now observing a moratorium on adding to their stockpiles of weapon-grade fissile material, but making this into a treaty obligation is widely seen as a crucial indicator of a preparedness to qualify and perhaps to abandon the view that possession of nuclear weapons is a core sovereign right. This is, after all, what has been asked, indeed often demanded, of those states without nuclear weapons. International support for this measure needs to be mustered so that any state disposed to muddy the waters with special interests or to simply engage in delaying tactics becomes aware that it could pay a significant diplomatic and political price for taking such actions. That said, it would be useful for Washington to consider providing stronger assurances on the limited goals of its BMD program, as uncertainty on this score feeds so directly into what other states regard as their minimum 'requirement' for offensive nuclear forces.

Coordinated international pressure should also be brought to bear on all states for markedly more transparency on plans and rationales for nuclear forces to counter instincts to hedge against uncertainties and allow political strategies for non-proliferation and nuclear diminution to be better informed and focused. To the same end, the accepted habit of the lesser nuclear powers to insist that they have no obligations to engage substantively in these processes until the big players cut back to parity with them should be vigorously contested. Similarly, any campaign to re-energise the non-proliferation enterprise must prevail upon Israel to declare its status as a nuclear weapon state and be subject to the same pressures and standards for transparency as the others.

The IAEA is already a strong advocate of precluding national production of fissile material in favour of internationally owned and operated facilities. As the case of Iran has highlighted, the right currently enshrined in the NPT to enrich uranium and breed plutonium for non-military purposes can only too readily tax the Agency's inspection rights and capacities to the point where it is unable to provide confident reassurance of compliance. The IAEA can be tasked to develop plans for a global network of such facilities that is responsive to projected demand, dispersed regionally to minimise transportation requirements and enhance security of supply, and which exploits any scope to co-opt existing national facilities. The costs are likely to be non-trivial, but they will still pale into insignificance against the prospect of the repeated use of force (the direct costs of the war in Iraq are now heading toward US$600 billion).

This has always been, and remains, a mammoth agenda, but in some crucial respects supportive changes are in prospect and should diminish the sensation of swimming so absolutely against the tide. In the first instance, there can be reasonable confidence that the United States will re-examine its approach to terrorism and recast some of the policy settings framed in the fusion of the 11 September 2001 attacks and the neo-conservative dominance in Washington. The United States will not step back from a determination to pre-empt any terrorist threat that it becomes aware of, nor from a determination to confront any state that offers safe haven to terrorist groups or, if it is a state with nuclear capacities, that threatens to transfer such capacities to terrorist actors. In the latter two instances recent precedents suggest that Washington would enjoy strong international support for such policy settings. Change is more likely in two respects. First, Washington will abandon the neo-conservative posture of overtly exercising and protecting pre-eminent power—a perspective that invested every position the United States took with an absolutist, unilateral quality. Second, we can expect to see a substantial recasting of the strategy for victory against terrorism in the context of this (1) being a long campaign with at best modest prospects for opportunities to bring decisive conventional military power to bear; and (2) therefore heavily reliant on relentless global vigilance to generate the requisite intelligence and to act upon it with police, immigration and customs resources as well as special forces and larger military capabilities on the (probably) rare occasions that this becomes appropriate.[6] We can expect to see in Washington renewed value being attached to reaching out and attracting international support, both because the harder, unilateral approach has delivered so little and because the US public has signalled that its appetite for the former can no longer be taken for granted.

The second instance of potentially swimming with the tide concerns America's nuclear force posture. As argued earlier, the Bush Administration seemed to go out of its way to stress that the target of some 2000 operational strategic warheads (and at least as many again in the immediate reserve) by 2012 constituted US minimal requirements for such forces. The position now emerging from Congress is that the development of a class of reliable, low-maintenance warheads to underpin US nuclear forces for the next several decades should be made dependent on a re-examination of nuclear strategy and related matters like the number and variety of weapons, the balance between operational and reserve weapons and the capacities of the nuclear weapons support complex to be retained by the United States. Just as individuals in Congress have stressed the consequences for nuclear non-proliferation of the United States mismanaging these issues, reviews of this kind can evaluate alternative ways of reconciling the continuing requirement for a timely nuclear capability and the stated objective of signalling a devaluing of and reduced reliance on these weapons.

Australia could commit itself to leading a like-minded coalition to take an agenda like this forward, with the intent to first rebuild the coherence of the nuclear non-proliferation regime and then to strengthen it further. This would no small undertaking, even given the energy that Australia routinely puts into this objective. It will require a significant investment of Australia's diplomatic and other resources, a determination to sustain this investment indefinitely, and a preparedness to encounter some tensions in relations with countries that are important to it. At the same time, as a wealthy non-nuclear weapon state allied to the United States, Australia is rather well-placed to do it in terms of capacity and credibility. Indeed, its activism in the arena of nuclear arms control during the 1980s in particular gives Australia something of a track record in the art of finding a path between fatalism and idealism. In addition, of course, we should forget neither the 40 per cent of the world's uranium, which we never wish to see returned as a mushroom cloud on Australians or anyone else, nor the fact that the further proliferation of nuclear weapons must eventually confront Australia with the decision on whether it can continue to live without its own bomb. Finally, if not Australia, then who?

Nor will it be an undertaking with any certainty of success. Perhaps the best that can be hoped for is a non-proliferation record that remains imperfect but manageable. We should therefore bear in mind that conceding that the non-proliferation battle has been lost is rather too likely to result in a record that we will see as profoundly regrettable.

Notes

1 For an illuminating account of this extraordinary saga, see Gordon Corera, *Shopping for Bombs: Nuclear Proliferation, Global Insecurity, and the Rise and Fall of the A.Q. Khan Network*, Scribe, Carlton North, Vic, 2006.

2 For a more extended treatment of these issues, see the author's, 'Missile Defence, the ABM Treaty and Nuclear Weapons—An Opportunity Missed', *Global Change, Peace and Security*, June 2002, pp. 87–104.

3 Statement by Assistant Secretary of State John S. Wolf to the Second Session of the Preparatory Committee for the 2005 Review Conference of the Parties to the Treaty on the Non-Proliferation of Nuclear Weapons, Geneva, 28 April 2003.

4 Ron Huisken, 'Uranium Sales to India: What Should Australia's Price Be?', *Defender*, vol. XXIII, no.1, pp. 21–22.

5 For one of the more recent explorations of the options in this field, see Harold H. Feiveson (ed.), *The Nuclear Turning Point: A Blueprint for Deep Cuts and De-alerting of Nuclear Weapons*, Brookings Institution Press, Washington, DC, 1999.

6 Since 2005, Washington has emphasised that its global 'war on terror' would be a long war and that conventional military power would have a comparatively limited role in prosecuting it. One has the impression, however, that this has been a tactical development, part of erecting a defensive political shield to deflect the backlash against a campaign that continues to go rather badly. It has not yet developed into a major reassessment of the terrorist phenomenon, positioning it more thoughtfully in the spectrum of challenges to stability and security, and cogently arguing for where the main thrust of the counter-terrorism effort should be directed.